SHALLOW-WATER DICTIONARY

Shallow Water Dictionary

A GROUNDING IN ESTUARY ENGLISH

John R. Stilgoe
with photographs by the author

PRINCETON ARCHITECTURAL PRESS

Published by Princeton Architectural Press
37 East 7th Street
New York, New York
10003
(212) 995-9620

ISBN 1-56898-029-9

97 96 95 94 4 3 2 1

First edition © 1990 Exact Change
ISBN 1-878972-02-2

Printed and bound in the United States by McNaughton & Gunn, Inc.

Cover design by Allison Saltzman
Book design by Naomi Yang

Library of Congress Cataloging-in-Publication Data
Stilgoe, John R., 1949–
Shallow-water dictionary : a grounding in estuary English / by John R. Stilgoe, with photographs by the author.
p. cm.
ISBN 1-56898-029-9 : $11.95
1. English language—Terms and phrases. 2. Salt marshes—Terminology. 3. Estuaries—Terminology. I. Title.
PE1689.S77 1994
427'.973'09169—dc20 94-22373
CIP

SHALLOW-WATER DICTIONARY

In mare multa latent. Oppian

In the Shallows the oarsman pulls precisely, for in the shallows a wrong stroke or two, a slight falling off before the wind, means a gentle grounding in clean white sand or grayish-black mud. Navigating the shoals and sandbars and negotiating the creeks meandering through almost limitless salt marsh require continuous picking along, constant choosing, countless corrections of course.

Here in the realm of the great marshes no sailboats, few engine boats, and fewer canoes intrude. Away from the main channel ripped open the day after Thanksgiving in 1898—in the storm which sank the steamship *Portland* and is still recalled by the name of its stateliest victim—the marshes enjoy an eerie peace created by the eternal nastiness of the "new mouth." Where the North and South Rivers—both estuaries—meet the sea in a surge of breakers and vicious currents that swamp boats and drown men (never women) is the newest geological feature of the North American eastern seaboard, a place government chart-makers label "strangers should not enter." Buoys warn of sandbars, but buoys drag, and even big-engined lobster boats remain moored on the "bad days" when wind and currents slam everything afloat into the surf thrusting beyond the buoys. Given an easterly wind pushing heavy rollers in from the Atlantic combined with an ebbing tide, the mouth proves impassable, and ferocious. Even on good days, those who approach it do so with dry mouths, respecting its reputation for instantaneous worsening.

So in the marshland still healing from the long-ago November shredding and rearranging, the gentle waters pass almost untenanted. Powerful engine boats explore the rare deep channels leading inland to marinas, but boats big enough to best the mouth on a good day draw too much water to probe the creeks, or drift upriver with the flow, under the disused drawbridges far into the wooded interior. Sailboats rarely enter, and almost never sail. Sandbanks and tiny islands allow no room for tacking, and movable centerboards prove almost as encumbering as deep keels. Canoes, seemingly ideal for shallow, serpentine channels and creeks, fare poorly. On a windless evening, between ebb and flow at high or low tide, they glide perfectly, but the paddlers study the weather, for in the world of the littoral, canoes demonstrate their fresh-water ancestry by swamping, capsizing, and spinning. Rowboats, anachronisms elsewhere in the maritime world of windsurfers, jet skis, and inflatable dinghies, seem to work.

Essay ghosts about the marshes. Sixteen feet long, offering three seats, a tiny decked-over storage locker in the bow (a dark place of spare oarlocks, first-aid kit, extra line, and emergency water bottle), and two seven-foot oars pivoting in bronze oarlocks, the bilge-bottom boat took shape from sailboat plans. Minus mast, centerboard, and rudder, and with seats rearranged for rowing, *Essay* carries an elderly two-horsepower Evinrude outboard engine, about as tiny and unreliable as engines come, but a perfect compromise for the long passages along the deep channels. Not powerful enough to push the boat against the currents in the bores, but light enough to be easily stowed amidships when *Essay* proceeds under oars, the engine runs almost silently. A pretty enough boat, then, with lines more gentle than a skiff, pushed by a diminutive, aging engine into sea-reaches where only oars give way, into the vast greenness where language proves as tricky as the channels.

Where the rower of *Essay* meets no other humans, language might be as unimportant as watches or clothes or credit cards. But the rower pulls with purposes other than exercise and recreation. He intends a book, a landscape history of the realm of estuary and marsh, and pulls as an explorer, as a chronicler of a place facing extinction as the earth warms and the sea level rises. Oriented toward the visual, he still needs to write, and in writing wonders about the quiet, almost mute vocabulary of his neighborhood, his environs. He wonders as he pulls along the creeks. In the quiet, the old language barely whispers.

This dictionary, a sort of salvage operation of words drifting from dictionary language, may serve to moor the terminology of estuary English. It is, however, mostly a cruise through an archipelago of half-sunken dictionaries, reference works touched with the drop-line of summer inquiry by a scholar intrigued with the death of topographical curiosity among the young, the young so absent from the marshes.

*

CREEK starts this dictionary of estuary English. From Maine to Cape Fear in North Carolina, the word designates, according to Hans Kurath in *A Word Geography of the Eastern United States*, "salt-water inlets of small streams that empty into the sea." Now *stream* is a general term for what New Englanders call a *brook*, and southerners call a *branch* or *creek*. Larger than a rill but smaller than a river, a brook is always a fresh-water flow. Early on, however, *creek* came to be a "corruption" of *brook*, as James Fenimore Cooper lamented in 1838 in his *American Democrat:* "'Creek,' a word that signifies an *inlet* of the sea, or of a lake, is misapplied to running streams, and frequently to the *outlets* of lakes." In his 1852 revised version of Noah Webster's *American Dictionary of the English Language*, Chauncey A. Goodrich understood *creek* to mean "a small inlet, bay, or cove; a recess in the shore of the sea, or of a river," but admitted too that it meant also "in some of the *American States*, a small river. This is contrary to English usage, and is not justified by etymology; but as streams often enter into creeks and small bays, or form them, the name has been extended to small streams in general." In his *Dictionary of the English Language* of 1859, Joseph E. Worcester, the great competitor of Webster, the champion of Harvard and Massachusetts English over that of Yale, Connecticut, and places west, defined the term as "a small inlet of the sea or of a river; a bay; a cove," but admitted it meant "in the Middle, Southern, and Western States" a "small river; a rivulet." But how small was small? In "The Witch of Wenham," Whittier writes "Of creeks that bore the salted sea/To pleasant inland farms," suggesting that creeks reached well away from the coast. Such was not the older sense, as the King James version of *Acts* 28:39 makes clear: "And when it was day, they knew not the land: but they discovered a certain creek with a shore, into the which they were minded, if it were possible, to thrust in the ship." What was the place "where two seas met" in which Paul wrecked? Surely not the creek of Whittier. By 1928, the Funk & Wagnalls *New Standard Dictionary of the English Language* reaffirmed the definition as "a small inlet" but refined the secondary one: "a tidal or valley stream between a brook and a river in size." Not until 1934, when the second edition of Webster's *New International Dictionary of the English Language* appeared, did the explorer of estuaries find something significant.

The 1934 Webster, perhaps the greatest of American dictionaries—and as the last hand-bound one, the last as thick as its editors wanted, not, as all subsequent machine-bound dictionaries, as thick as the biggest binding machine in the Republic could bind it—explains *creek*, in part, as follows: "a small inlet or bay, narrower and extending farther into the land than a cove; a narrow recess in the shore of the sea, a river, or a lake...the estuary of a small river or a brook, emptying on a low coast or into the lower reaches of a wide river, together with the upper course of the small river or brook to its source." Here, after centuries, some sense, some awareness of the low coast, some precision about the estuaries of brooks. But not enough.

In the marshes a creek may be the estuary of a brook, although not always. Satuit Brook empties directly into Scituate Harbor. No part of it any native calls a creek. But a mile away, across a cornfield last flooded by spring tides in 1786, creeks reach across marsh from creek to creek, lacking all connection to inland brooks. In them *Essay* moves slowly, under oars, and sometimes, above an oar forced into the muddy bottom when the creek is too narrow for rowing. Such creeks are cracks. Not only do they appear so from atop the marsh, when they stymie hikers, hunters, and fishermen caught by rising tides, they appear so from the boat. Always the banks are caving, the saltmeadow cordgrass and creek sedge grass losing their grip on the mud and tumbling into the current. The marsh is cracking always, always quietly, with only a splash now and then to mark the fall of bank.

How far up a creek does a brook begin? Where salt water turns brackish? To crack the marsh realm, to burgle it by boat, involves naming such indistinct zones as the creek-brook boundary. Words like *inland* advertise the land-lubberliness of dictionary editors. No antonyms exist, not *outland.* Maybe *seaward* serves here. Creek ends slightly seaward of brackish water. Creek ends with the zone of saltmarsh plants demanding salt water. Where fresh-water plants reign in coastal New England only *brook* serves to designate the flowing water—unless *Essay* creeps up Satuit Brook, salt at least a quarter mile along its brookness.

Then too, a creek always has a muddy or sandy bottom, never a rocky one. Brooks flow over stones. Creeks flow—with the tide—over soft, squishy stuff, stuff that sucks off loafers and low sneakers, stuff that shelters razor-sharp broken shells, old fishhooks, and long-sunk aluminum pop-tops, things that slice bare feet. Part of any saltmarsh definition of *creek* must include the sucking, never rocky bottom that swallows feet, knees, and thighs, and unabridged dictionaries too, for that matter.

Creek cruisers lack SEA-MARKS, at least as estuary English defines *sea-marks.* The 1852 Webster defines the term as "any elevated object on land which serves for a direction to mariners in entering a harbor, or in sailing along or approaching a coast; a beacon; as a light-house, a mountain, etc.," while the 1859

Worcester explains it as "a point of land, or an object on land, as a light-house, to assist mariners in directing their course." Buoys certainly assist mariners, but BUOY is a word which receives scant attention in the 1859 Worcester but rather more in the 1852 Webster, wherein it is defined, in part, as a "floating mark." A little rummaging makes certain that a sea-mark is always on land, but thereafter may be well-nigh anything, including the land—even a mountain. But the British, at least in Cornwall, use the term to identify the utmost reach of the tide—the point where creek becomes brook, perhaps. In New England salt marshes the word long meant particular objects on land useful in conjunction *with each other* for purposes of alongshore navigation—Justin Winsor, in his 1849 *History of the Town of Duxbury*, used it so: to find good fishing ground "bring Indian Hill just over the point of Manomet, and Captain's Hill over the south part of the Plumb Hills on the beach." The 1934 Webster offers two meanings, "a line on a coast marking the tidal limit; hence *full seamark*" and "an elevated object discernible at or from sea and serving as to guide or warn mariners; a beacon; a landmark; hence a sign of danger." A landmark, therefore, *can* be a sea-mark, but does the 1934 Webster really assume equality? Is a sea-mark always visible from land?

Where *Essay* pokes its bow, a sea-mark stands always on land (unless a storm has wrenched it adrift) and always elevated (unless it has toppled over, say from the weight of a gull). In the marsh-world, a sea-mark is a stick jammed in the marsh at the edge of a creek. If the creek is deep and wide and populated by summer people in engine boats, marina-owners may nail green and red arrows (sometimes with reflectors) to the sticks, warning the fair-weather navigators away from mudbanks. But if the creek is narrow and rarely traveled, the sea-mark is often a sapling shoved into the marsh, or an old lobster-pot lath, sometimes topped with an inverted beer can or plastic bleach bottle.

What use are the least formal sea-marks? At dead low tide, when the creek is almost empty of water, the bank towers five feet above the rowers, and the sea-mark towers three feet above the top of the bank; at full tide the mark is easily touched from the passing boat. Why mark the narrow creeks at all? Are the marks like the fishing poles Thoreau remarks in *A Week on the Concord and Merrimack Rivers*, some left-behind totem of luck? No. The sea-marks are landmarks. Where all creeks look similar—and where mudslides and marsh-slides change appearances constantly, the sea-marks help the mariner determine his path—not his course—through the marshes. Row further and turn left—only summer people use "port" and "starboard" in the marshes—at the mark, row another ten minutes and find the main channel. But as the tide ebbs, the rower must leave his boat to gain the top of the marsh. He must clamber up onto the land to discern the sea-mark. Who sets sea-marks? No one knows. No one knows who moves them, either.

Once adrift on the tide, a sea mark is FLOTSAM, and once stranded by the ebb, it is JETSAM, left on the WRACK LINE. Dictionaries other than this define *flotsam* as pieces of a wrecked ship or its cargo floating on the sea—not just anything floating around, say trees washed south from Nova Scotia—and *jetsam* as stuff deliberately thrown from a ship in danger of foundering—jettisoned—and, Worcester is specific about this, floating under the surface of the water. Jetsam attached to a floating buoy for eventual recovery is *ligan*, but

At sundown, light and tide ebbing together, a creek withers into gutter, ending in unnavigable mud.

about that nothing is known along the marshes, so flotsam and jetsam receive attention here, along with wrack. The 1934 Webster is clear that *wrack* designates any wreckage *and* all cast-up seaweed, the last usually remaining in a black line along a beach. In the marshes no one uses *flotsam* and *jetsam* anywhere near accurately, and *wrack* designates chiefly stranded seaweed, although the word more frequently identifies a wretchedly maintained boat "going to wrack and ruin." Does a correct term exist for debris, floating or stranded? Is the telephone pole long ago flung across the marsh and into a creek jetsam or wrack or what—aside from an impediment to navigation?

How does one talk—or write—precisely about a place in which wandering mariners climb ashore to seek out sea-marks, where *flotsam* identifies everything afloat not owned? What use is it to trace *skiff* far back to the German *schiff*, the word now translated as *ship*, or read in John G. Rogers' 1985 *Origins of Sea Terms* a definition of skiff: "a small boat, now flat-bottomed"? Is estuary English the gentle echo of older speech or the raw edge of a new tongue?

SKIFF. The 1852 Webster calls it "a small, light boat, resembling a yawl," while the Worcester defines it as "a small, light boat; a wherry," and the 1934 Webster, in the midst of defining two rare usages (both sailing vessels), focuses on "a light rowboat." The 1928 *New Standard Dictionary* provides "usually, a small light boat propelled by oars or sculls; formerly, a small sloop-like vessel." But the *American Encyclopedic Dictionary* knew better as early as 1894, explaining a skiff as "a small flat-bottomed boat, usually without a keel; a small boat generally. In some parts of the United States the term is applied to a small boat, sharp at both ends, generally having a centerboard, and provided with a small triangular sail." In the broad marshes dividing Scituate from Marshfield, in the estuary linking the sea with the towns of Norwell, Pembroke, and Hanover, in the great bays of Duxbury and Plymouth just south, *skiff* designates a flat-bottomed boat, usually under fourteen feet long, and propelled by oars or outboard engine, and pointed—"sharp" in the terminology of the *American Encyclopedic Dictionary*—at one end only. Other boats of similar size are not skiffs. The rare flat-bottomed boat blunt on each end was known into the 1960s as a PUNT and into the early 1970s as a DUCKBOAT, although the terms, like the boats, were rare indeed, even in hunting season. Catalogues and mass-marketing programs injected the southern JOHNBOAT into marsh speech, but now that term designates something else, an open, fiberglass boat with one pointed end, propelled by an outboard engine and having two elevated swivel chairs suitable for sport fishing. In a good chop the elevated chairs magnify the hull motion, and few johnboats remain long in salt water—their owners trailer them back inland, where they belong.

Skiff has come and gone and come again—tide-like—in English. Skiffs were in at the founding of the Massachusetts Bay Colony, when John Winthrop noted in his diary, in September, 1630, that he watched the master of the *Arabella* order "his skiff to be heaved out" to take him across to a strange ship encountered by the fleet bound for New England. *Skiff* deserves more than shallow-water sounding. It demands a bit of blue-water notice. In 1702 John Kersey, in his *New English Dictionary,* understood the small-boatness of *skiff.* "A ship-boat," he called it, perhaps the boat most quickly heaved out on matters of business. In his preface, Kersey explained his desire to define especially "useful terms in all faculties," among them "divinity, ecclesiastical affairs, the statute laws of this nation, history, geography, maritime affairs, plants, gardening, husbandry, mechanicks, handicrafts, hunting, fowling, fishing, etc." Unlike his lexicographer predecessors, Kersey aimed largely at defining—"explaining" in his words—common speech, the essential words of everyday speech, not the language of the literati. Kersey wanted a down-to-earth—and down-to-sea—dictionary, and railed against earlier ones intended for the erudite, dictionaries so grounded in ancient languages that "a plain country-man, in looking for a common *English* word, amidst so vast a wood of such as are above the reach of his capacity, must needs lose the sight of it, and be extremely discouraged, if not forced to give over the search." Kersey thought *skiff* an important term, and took pains to explain it, however briefly by later standards.

Between Winthrop and Kersey lies less than a century, but a century in which one dictionary-maker tried translating *skiff* into "the language of the natives in that part of America, called New-England." In his *Key into the Language of America* of 1643, Roger Williams not only explained the Narragansett technique of burning a canoe out of a log, but something of the Indian understanding of English rowing boats. *Mishoonemese* designated "a little canoe," and *wunnauanounuck* identified a shallop, the big rowing-sailing boat used for exploration by the settlers of Plymouth, but *wunnauanounuckquese* meant "a skiffe." The suffix suggests that the tribesmen understood the skiff to be smaller than a shallop, something the Plymouth colonists ought to have considered before venturing from the *Mayflower*, but more of that later.

A skiff rows harder than a round- or bilge-bottomed boat, but perhaps it floats in a little less water. Children row skiffs, often learning to loathe rowing. Why skiffs became so popular, and remained so into the 1960s, when fiberglass dinghies, then inflatable boats replaced them, requires an investigation of sheet plywood as a post-war boat-building material. A four-by-ten foot sheet of plywood proved the perfect raw material for the flat bottom of the skiff, and two more sheets provided the two sides. But for a skiff to perform well under oars, advises the retired builder of hundreds, it must have a SKEG.

Of the word *skeg* the old Webster and Worcester remark only that it designates a "wild plum"—the Worcester adding also "a wild oat." But the *American Encyclopedic Dictionary* and the *New Standard Dictionary* note that it identifies also a triangular piece of wood fixed under the stern of a flat-bottomed boat. The 1934 Webster is quite precise about it too, adding that the piece of wood "takes the place of the afterpart of the keel." As *skiff* came to designate not a sort of sailing vessel but a flat-bottomed rowboat, *skeg* received increasing attention from lexicographers.

And well it should. Without a skeg a skiff is vicious to row. A skeg-less skiff wanders about, drifts with the wind, and quickly exhausts the already infuriated rower. A good skeg reaches perhaps six inches below the bottom of the skiff, and as a fixed rudder (or aftermost part of an otherwise non-existent keel) directs the boat in a straight line.

Of course, a skeg adds draft to a skiff. Add a nice deep skeg to a skiff and it draws a half foot more water, and runs around that much more frequently. Moreover, a skeg cants a grounded skiff, and if the skiff grounds on rocks or hard sand, the skeg often twists, then splits, and becomes flotsam. To skeg or not to skeg bedevils those who wish to "skiff" the marches—to use a verb Shakespeare used, although infrequently—and often the unwise forego skegs. *Essay* pulls well, carrying an unusually long skeg, but one only four inches deep, not much on a sixteen-foot bilge-bottom (a hull shaped between flat and round) boat, but enough to keep the vessel gliding on a straight course while the historian rests on his oars.

Resting on oars demands good oars, oars well balanced to begin with, and secure in locks. OARS ought to have vastly more attention than they receive. Why do environmentalists shun them, unless environmentalists adore canoes so much that paddles blind them to oars? And oars confuse. Hear Henry James, confusing matters in 1909 in *The Ambassadors:* "What he saw was exactly the right thing—a boat advancing round the bend and containing a man who held the paddles and a lady, at the stern, with a pink parasol." Not the right thing at all. Paddles belong in a canoe, not in a boat.

An oar has parts. Before the era of outboard engines, skiffs and other small craft moved "under oars," at least on windless days. The "white-ash breeze," as old-timers still call it, took its name from the wood from which good quality oars come, although spruce is replacing white ash. Who bothered defining *oar* in a time when every one understood the term, when even Richard Henry Dana, in his 1851 opus, *The Seaman's Friend ... A Dictionary of Sea Terms,* noted only that an oar is "a long wooden instrument with a flat blade at one end, used for propelling boats." Did landsmen — James, for one — know any more, or even need to? The old Webster says little more, "an instrument for rowing boats, being a piece of timber round or square at one end, and flat at the other. The round end is the handle, and the flat end the blade," and the contemporaneous Worcester nothing different. But more makes an oar than handle and blade. After all, an oar is an "instrument," not a tool, something for precise work, something whose rhythm approaches music. Its form has long perplexed dictionary-makers. Between handle and blade, Webster's "piece of timber," is the shaft, made round with a spoke-shave. But between the handle and the oar-lock, the bronze, pivoting horns in which the oar plays, is that part of the shaft known in the estuaries as the LOOM. Although the 1914 *Century Dictionary: An Encyclopedic Lexicon of the English Language* defines loom as "the part of an oar between the blade and the handle; the shaft," the estuary definition emphasizes the importance of one part of the shaft.

The loom gives grace to the whole oar. Thicker than the rest of the shaft, its weight counterbalances the part of the oar outboard of the boat. Properly shaped, the whole oar becomes sculpture, almost weightless in its balance, almost effortless to PULL. Novices ROW a small boat, but those seasoned in the estuaries pull. Landsmen paddle when their outboard engine dies, hunching in the bow, clutching a sporting-goods-store canoe paddle, lunging at the waves. Novices, assuming a graceful little boat, perhaps a skiff with skeg, and well-balanced oars, discover almost immediately the joy of pulling the boat "under" the oars. A little longer in a bigger boat moving under two pair of oars, and the novice learns the ancient meaning of INTERFERE, "to put in one's oar," often in the wrong place. But a little practice suffices to teach novices how to pull in harmony, and the synchronized rowers glide remarkably fast, except through bores.

BORE now designates a talking nuisance. "The spell of a symbol is broken when a bore explains what it means," asserts H. M. Tomlinson in *A Mingled Yarn*. But in the marshes, the word identifies the mad, noisy rush of the tides *against the current* in a narrow channel. The old Worcester makes it clear enough: "a rapid and noisy influx of the tide into a river or strait against a current." Of course, the word also may indicate the place where the tidal action occurs, suggesting that the constriction is something like a bored hole, something like the bore of a cannon. In the precise sense of the 1934 Webster, a bore is "a tidal flood which regularly or occasionally rushes with a roaring noise into certain rivers of peculiar configuration or location, in one or more waves which present a very abrupt front of considerable height, dangerous to shipping." And after listing some rivers, among them the Amazon, the same Webster continues: "less properly, a very high and rapid tidal flow, when not so abrupt, such as occurs in the Bay of Fundy and in the Severn." But in the marshes, contemporary vernacular speech veers from such precise definition.

Depending on the amount of fresh water pouring seaward and the strength of the flood tide (and easterly wind), the bore varies from scarcely noticeable, merely a swift current, to a surge. But "the bore" is a place, the frighteningly narrow channel of the North River between the rotting abutments that once held a long-dismantled railroad bridge. The railroad embankment, built long before the days of environmental impact statements, still streaks across the marsh, now serving as a straight-line nature walk sliced into sections by the tide. At the bore the tide can race so strongly that small outboard engines cannot best it, and large, heavy-engined craft find their passage in doubt. No one rows against the bore.

Today *bore* usually connotes something of the tedium of boring a hole with an auger, or borer, but to enter the North River bore when the tide and current slam each other precisely between the abutments means staking navigation, and life, on the steady roar of the engine. Old-timers pronounce *bore* almost in two syllables, "bo-ah," accidentally reiterating its Old Norse root, *bara*, "a wave." On a Saturday, with large and small boats barely steerable and jammed together like so much flotsam, the place shames contemporary inland uses of the word. One can be bored to death.

Why do words drift so far from their original meanings? Emerson wondered in *Nature*, concluding that "words are signs of natural facts," but perhaps now dictionaries too completely sever readers and talkers from the enterprise of dictionary making. In the 1860s, educated people rummaged in the origins of American words, words like *schooner*. In June, 1862, one anonymous letter-writer asked the readers of *The Historical Magazine* for information about the word: "This word is said to be of American origin. (See Worcester's *Dictionary*.) It is not to be found in the dictionaries of Bailey, Johnson, or Walker. Can any reader of the *Magazine* furnish a quotation from any old book or newspaper, American or English, which will serve to fix the land of its birth, or the date at or about which it made its appearance in the language? Is the word *schoon* (from Anglo-Saxon *scunian?*) in present use among the common people in any part of the United States?" Such questions suggest a serious inquiry in progress, and a bit of sophistication about the language of the "common people," what is, to be precise, the "vulgate." The 1852 Webster only defines the term, far more briefly than Dana does in his 1851 *Dictionary of Sea Terms*. "A small vessel with two masts and no tops," even in a maritime age, says little. But Worcester found the origin of the term almost enchanting. His *Dictionary* relates the story of the launching of the first such craft in Gloucester around 1714. As the vessel slid into the water a bystander cried, "O see how she schoons," and her owner shouted, "A schooner let her be." A pleasant story. Within a month of the anonymous query, *The Historical Magazine* published a lengthy reply, citing the 1757 use of the term in Diderot's encyclopedia, in a subsequent British encyclopedia, and in New York colonial records. The next month appeared two more replies, providing full documentation of the Gloucester tale. Genuine interest quickened the writers. They enjoyed tracing words, rowing into the creeks of maritime American speech.

A mendicant guide at best, the seamark has a nasty way of playing windvane in a gale, then seducing the rower into sandbanks or worse.

Schooners rarely move in the estuaries now, although some tiny sailing vessels are still "schooner rigged." Is the word American, or derived from some Scottish term referring to skipping flat stones on the sea? The 1966 *Random House Dic-*

tionary of the English Language provides nautical definitions, a drawing, a glance at prairie schooners, and a second definition: "a very tall glass, as for beer." What about the second definition? When did that usage enter American speech? And why, if a schooner has "no tops," is the glass tall? Long before *skyscraper* designated a tall building, it identified the top-most triangular sail on a full-rigged ship. Where "a schooner of beer" came from now puzzles as *schooner* did in the 1860s, but how does the rower ask for help? *Essay* carries neither masts nor sails, and no beer either, and *schooner* is a deep-water word anyway and best avoided—unlike GUZZLE.

Of all the words in the marshes, *guzzle* is the most luscious, the most demonstrative of the enduring qualities of spoken language, and now, perhaps, most nearly an echo. Guzzles come and go. As transient features in barrier-beach topography, they exist at the whim of storms, especially winter storms. Almost never defined, the term lives in the marshes as a transplant from Kentish marshland dialect. The 1934 Webster calls it "a gutter or drain," and notes the word is obsolete, except in English dialect. Yet it lingers in the estuaries as obsolescent, not obsolete, for it has a use. Even as the Inuit have a dozen words for snow of different sorts, the natives along the estuaries need *guzzle*, for they need to talk about the low spaces on barrier beaches that sometimes allow spring tides or storm tides to pass over the beach into the marshes inland from the sea. In *Cape Cod* Thoreau speaks of "bars and guzzles," but what dictionary now unlocks his words—and what he saw to write them?

Summer residents walk the sandy beaches and notice low spots, places where the dunes are low, young, and especially fragile, and squish their feet in the damp sand crossing the dry.

Unknowingly — unspeakingly — they cross a guzzle. Dry at low tide, dry at almost every high tide, the guzzle is a sort of way for the ocean, a potential crossing-through place for the sea. Years pass without the breaching, the dunes rise, and perhaps the guzzle is no more — but perhaps the sands beneath the dune are weak, and the guzzle splits open in some spring gale.

In 1898 a guzzle split so widely that the estuaries had a new opening to the sea, the "mouth" of the North River as locals call it, the "New Inlet" in the words of government cartographers. Since the settlement of the marshes by Kentish immigrants in 1633, guzzles have split open and filled up, usually about one every eighty years. Indian place names suggest the process is ancient. Musquashcut Pond, more or less fresh water when settled by the English, was once an estuary, for that is the significance of the suffix "cut," the same suffix the ends "Connecticut." But not for decades did the sea come through the guzzle and flood the pond and surrounding cornfields. *Guzzle* identifies a sort of gutter or drain, then, one almost invariably dry, the sort of landform Kentish people knew, and distrusted. Now the word is faint, for the guzzles are high. Not since the late eighteenth century has the sea crossed one cornfield, and the trees around it grow old and gnarled in the constant sea wind. Only old timers — and landscape historians — wait for what the experts now call a "century storm," the storm that has a one per cent chance of happening any year, the storm the old-timers say, dismissing the one-per-cent experts, comes once a century.

Storms like hurricanes, the Portland Storm, the Blizzard of 1978, and other Acts of God (as insurance adjusters call them) slam through guzzles and course inland along GUTTERS. In the littoral where *Essay* moves so silently a gutter has nothing about it of paved streets and curbstones. A gutter is a dug course for water, not natural and winding as is a creek, but usually straight. Now called a DITCH, most marsh gutters date from the Depression, when the Civilian Conservation Corps dug hundreds to drain the marshes after every high tide, and so lessen the stagnant pools in which mosquitoes hatch. Two feet wide and six (or less) deep, the gutters trap the unwary hunter or hiker, for the marsh grass bends over them, totally masking them. About *gutter* most dictionaries provide only confusion; even the 1934 Webster focuses on draining. But *gutter* once meant something more, in a way more sinister than it, or *creek*, does now.

Thoreau crossed a gutter, and got his English mixed. "We crossed a brook, not more than fourteen rods long, between Orleans and Eastham, called Jeremiah's gutter," he noted in 1865 in *Cape Cod*. "The Atlantic is said sometimes to meet the Bay here, and isolate the northern part of the Cape." Now "brook" is not the precise term, for a gutter is not a brook, and anyway, the gutter had become a guzzle. In 1717, Cyprian Southack rowed a boat across Cape Cod along the gutter Thoreau so easily forded. The gutter, dug it seems by someone named Jeremiah, was widening with the scour of the tides by the time Southack and his men rowed through it to surprise pirates at anchor at the opposite end. Prudent Cape Codders later filled it, to keep the outer Cape from becoming an island, just as they did the guzzle between Truro and Provincetown which Thoreau saw filled with sods, brush, and planted beach grass.

Southack's surprise passage suggests what the 1914 Century makes clear in its definition of *gutter*. In its thinking, a gutter is "a passageway; a secret passage." At least one gutter remains a well-kept secret in the marshes now—the gutter linking the Marshfield marshes with those of Duxbury. Only kayaks and canoes, and then only at spring tides, can navigate it. *Essay* draws too much water, perhaps, but has not tried to force a passage. But what an inland row, along that gutter meandering behind the barrier beaches, that might be. Far better than motoring through a GUT, the most unsecret of straits, like Hull Gut fit chiefly for warships and freighters. No, a gutter remains as Chaucer used it, a secret passage, not natural but made, made by some farmer for draining his salt-hay fields, or to give access to a GUNDALOW.

A gundalow is, or rather was, for not one exists today, a flat-bottomed, blunt-ended boat used for moving hay from the outer marshes to the docks of farmers, "a lighter or large flat-bottomed boat on the rivers of New England," in the slightly vague words of the Century. Some were big enough to freight horses and small wagons, the horses wearing massive wooden shoes strapped to their hooves to keep them from sinking in the mud, and some gundalows, if one believes Whittier's poem "Snow-Bound," sailed:

> We heard the tales of witchcraft old,
> And dream and sign and marvel told
> To sleepy listeners as they lay
> Stretched idly on the salted hay,
> Adrift along the winding shores,
> When favoring breezes deigned to blow
> The square sail of the gundelow
> And idle lay the useless oars.

In the North and South River marshes, however, few if any gundalows sailed—the awkward vessels floated up and down the twisted estuaries with the tides, guided by long oars called sweeps and shoved by farmers leaning on poles. Drifting haystacks, carrying the hay stacked earlier in the day on STADDLES, poles set into the marsh to keep the stacked hay from tidal wetting, gundalows are now memory, and, like the ruins of staddles preserved by salt water from rot, their ribs, or rather the ribs of one, thrust up through the marsh grass, reminding the rower and walker that wagons and pungs once went to sea, or at least across the flats and spits.

FLATS and SPITS deserve notice, and not only because careless mariners ground on them and spend the whole day waiting for another tide. Flats nearly defeated the Plymouth colonists, and certainly stymied their explorations of the coasts of Cape Cod and what they called Plymouth. As Bradford remarks, they "had much ado to put ashore anywhere, it lay so full of flats." The shallop did badly in shoal water, unlike a skiff. Useful for clamming—what Cape Cod locals call SCRATCHING from the marks made by the clam rake—flats are not attractive in the eyes of tourists. Marshes they like, and the open sea and the bays, but flats exposed at low tide stymie them as they stymied Bradford. Four-wheel-drive vehicles sink in them, no boats cross them, even walking is hard, and sometimes impossible. Walking across the sucking, grasping mud of the flats is harder than climbing hills and mountains, for shoes and boots are pulled off and bare feet encounter broken clam shells and shards of glass. The Random House dismisses flats—"a marsh, shoal, or shallow"—but the Worcester knows better: "a place under water of but little depth; a sand-bank or bar; a shallow; a strand; a shoal." A more solid definition, and one suggesting the varieties of flats a mariner or landsman encounters. The 1934 Webster adds "alternately covered and left bare by the tide," and the combination of definitions suggests the difficulties of explaining land that is sometimes under water.

Where flats meet the water at low tide may be a spit, what the Century calls "a small point of land running into the sea, or a long narrow shoal extending from the shore into the sea." Spits torment summer people. Far beyond the flats, gleam the spits, their white sand contrasting with the grays and browns of muddy flats. Utterly inaccessible to walkers, especially to those who wonder about quicksand, the spits are the private beaches of the small boaters, those happy few able to navigate the creeks cutting the marsh and flats, and able to pull their craft onto the hard sand. Away from dune buggies and crowds, the boaters picnic on what dry sand remains untouched at high tide, rewarding themselves for a stiff pull or gentle cruise, and amusing themselves by watching the big powerboats aground on the shoals far away from the spit. Spits tend to be inaccessible.

Solid, seaworthy definitions of alongshore things tend to be inaccessible too, now that unabridged dictionaries offer only the briefest of explanations, and sometimes none at all.

*

Scrutinizing requires discrimination, and discrimination must be learned. And the learning proceeds with words, whatever the so-called visualists say. To scrutinize the landscape—or seascape—of the marshes requires some discrimination of its components, some telling of sea-marks from flotsam, guzzles from gutters. A precise word here and there makes the components not only more easily discriminated, but perhaps more easily remembered.

Yet the marshes are almost invariably deserted, at least away from the major channels. People tend to find them uninteresting, not variegated enough, monotonous, boring, although finding out what such people think, when such people avoid the marshes, proves a little difficult. Exploring in what is at first glance a seamless web of green undoubtedly pales beside roaring about the open ocean in an engine boat or gallivanting around the crowded yacht harbor a few miles away, or so it seems. And anyway, who pulls a boat today?

Start there. Pulling is out of fashion. Until it is learned it makes for hard work, and the outboard engine produces far more speed. Pulling, like walking, requires a certain cast of mind not only out of fashion, but nearly beyond the reaches of imagination. In 1750, five years before he published his *Dictionary of the English Language,* Samuel Johnson asserted that "it ought to be the endeavor of every man to derive his reflections from the objects about him; for it is to no purpose that he alters his position, if his attention continues fixed to the same point." In 1862, while *Historical Magazine* readers tracked the origin of *schooner,* Thoreau argued, in an essay entitled "Walking," that he had met almost no one "who understood the art of Walking, that is, of taking walks,—who had a genius, so to speak, for *sauntering:* which word is beautifully derived 'from idle people who roved about the country, in the Middle Ages, and asked charity, under pretence of going *à la Sainte Terre,*' to the Holy Land" [Thoreau quotes Worcester here, although without acknowledging his source—and Worcester took his derivation from Johnson's *Dictionary*]. Johnson and Thoreau suggested that most people essentially ignore their surroundings, and walk, if they walk at all, oblivious to nearly everything. Their thinking has been restated, frequently. "To a person uninstructed in natural history, a country or seaside stroll is a walk through a gallery filled with wonderful works of art, nine-tenths of which have their faces turned to the wall," asserted Thomas Henry Huxley. Such is the faith of all who write field guides, who direct nature walks, who attempt to teach others to discriminate. But field guides depend on words, and words require definition.

"Burton's *Anatomy of Melancholy* is a book of great learning. To read it is like reading in a dictionary," Emerson remarks in his "Books." "Neither is a dictionary a bad book to read. There is no cant in it, no excess of explanation, and it is full of suggestion,—the raw material of possible poems and histories. Nothing is wanting but a little shuffling, sorting, ligature, and cartilage." At mid-nineteenth century, Emerson's faith in dictionaries was justified. The Republic, bent on creating a national literature, hoped to strengthen its national language—thus the intense interest in dictionaries, the competition of Webster and Worcester, the searching out of derivations. But now, in the late afternoon of the twentieth century, dictionaries are so much thinner, literally, than they were in 1914 when the Century made its massive appearance bound in corduroy, or in 1934, when the Webster arrived in bulk enough to enrich the Great Depression.

So little valued are old unabridged dictionaries that sellers of used books throw them away. Most people have no desire to own an unabridged dictionary, and many, even college students, use their abridged versions chiefly for checking spelling. And slowly, and as surely as the coming of the tide, the computer software that "checks spelling" threatens the last general use of the light-duty dictionary. The writer of dictionaries, argues Johnson in the preface to his great work, is "doomed only to remove rubbish and clear obstructions from the paths through which Learning and Genius press forward to conquest and glory, without bestowing a smile on the humble drudge that facilitates their progress." Johnson, Webster, Worcester, Funk and Wagnalls, the editors of the Century and Random House dictionaries, and the others carry an awesome responsibility, however. They must clear away the rubbish and obstructions well, lest learning and genius go astray. In the speech of the marshes, definitions are sea-marks, and a wrong one here and there can turn a pleasant cruise into a very real crisis, say on a late autumn afternoon with a rising wind and *skift* of snow in the air.

Now that aircraft fly the oceans, the edge of the land receives little attention from the humble drudges. What does the intrigued student of Thoreau's *Cape Cod* make of the expression "bars and guzzles"? In what contemporary dictionary does he or she find *guzzle* defined, let alone fully defined, with its connotation of the sea guzzling the land —or the land the sea—the way a drunkard guzzles beer? Where does the inquirer find any definition of Whittier's oddly spelled *gundelow*? Who knows now that a gutter is a secret passage?

*

Estuary English, the marshland vocabulary, whatever its proper name, endures out of need. The fisherman, the hunter, the boater, all need to designate guzzles and creeks, need to speak precisely about constituents of a landscape of importance. Perhaps the terminology is obscure, and certainly it is chiefly oral, as surely as it is old. "Obsolete words are admit-

Here and there a bit of high ground, usually covered in bayberry and beach rose, marks a refuge from all but the highest spring tides.

ted," announced Johnson in the preface to his *Dictionary*, "when they are found in authors not obsolete, or when they have any force or beauty that may deserve revival." Estuary English endures in authors not obsolete—everyone slightly intrigued by the natural realm reads Thoreau—and has force, an energy lacking in much standard terminology.

Visualists, especially painters, poets, and historians of landscape, suffer from the fading of such specialized vocabularies. Thoreau marveled at the colors of a Cape Cod autumn, at the "incredibly bright red of the Huckleberry, and the reddish brown of the Bayberry, mingled with the bright and living green of small Pitch-Pines, and also the duller green of the Bayberry, Boxberry, and Plum, the yellowish green of the Shrub-oaks, and the various golden and yellow and fawn-colored tints of the Birch and Maple and Aspen." But how do the explorers of the marsh note colors? Do they agree with Thoreau that the colors of the marsh are "warmer colors than I had associated with the New England coast?" Are the colors softer? What *is* the yellowish green of the shrub-oaks? Chartreuse?

Dictionaries do badly on color definition nowadays. Once they assumed educated, inquiring readers. The 1934 Webster defines *manila* as "a color, yellowish red-yellow in hue, of low saturation and high brilliance." In the marshes *manila* identifies old-fashioned line, or rope, not nylon. Is the line, made from manila hemp imported from the Philippines, manila-colored, or is "manila paper," of grade-school arithmetic exercises, closer in hue and saturation? What modern dictionary defines *manila* as a color, a color equal in importance with old landscape colors like umber and sienna?

CHARTREUSE is a splendid word to crash into during a marshland voyage, a word linked somehow, perhaps by a gutter, to CHART. The root word is old, echoing across an abyss of time from the English *card* to the Latin *charta* meaning paper, itself derived from the Greek *chartes*, somehow meaning "a leaf of paper," a word the Greeks took from Egypt, the home of paper, although the papyrus plant is a clearer source for the contemporary "paper." Books still have leaves, but no longer does CARD designate a chart. Now and then into the estuaries chugs a yacht, however, its compass needle floating above the "compass card"—so a whisper of *card* as meaning a very important leaf of paper survives. Chartreuse survives too, but as a word nearly bereft of origin, a piece of philological flotsam streaked with "charter" and "charterhouse"—and "charthouse."

Worcester knew *chartreuse* only as "from the desert of Chartreuse, in Grenoble, where the order of Carthusians was established by Bruno in 1086," and the 1852 Webster told inquirers little more: "a celebrated monastery of Carthusians, in the mountains of Dauphiny, in France. The regulations are exceedingly severe." But the regulations apparently permitted the distilling of liquor. By 1934, the inquirer could find a great deal about monastery activities merely by looking into the unabridged Webster. Beyond the first definition, focused on the site of the monastery, are others: "a liqueur of complex composition, made by Carthusian monks, esp. formerly at La Grande Chartreuse. Balm leaves, orange peel, hyssop, peppermint, etc., and various spices are said to be used in its manufacture. Of the three qualities, green, yellow, and white, the first is the most valued and the yellow is the sweetest." But the inquirer learned more: "a color, yellow in hue, of medium saturation and high brilliance. It is less red and of higher saturation and brilliance than canary yellow." And if the inquirer read further, he or she found separate definitions of *chartreuse green* and *chartreuse yellow*. In afteryears, however, lexicographers lost their grip on the gradations of meaning. By 1939, *The Winston Dictionary: The Advanced Edition* offered only "a pale green or yellow tonic liqueur, originally made by the monks at the first Carthusian monastery; a pale green color." Twenty-eight years later, while noting that the liqueur was made "at one time, at Tarragona, Spain," the Random House *Dictionary* told searchers that the word designates "a clear, light green with a yellowish tinge." So much for the complex definitions of the 1934 Webster.

In these last years of the century, and of the millenium, charting the definitions of *chartreuse* becomes anachronistic, like pulling a boat against wind and tide, like using manila rather than nylon line. Color photography replaces Thoreau's concern with describing colors, does it not?

The Eastman Kodak Company thinks not. In 1985, in Reference Sheet E-73, it warned photographers of hazards in their way, setting up a sort of sea-mark for the cautious, for those photographers willing to read. "Other colors—such as shades of chartreuse, lime, pink, and orange—may not reproduce so well. It would be possible to design a film that would reproduce these colors better, but then some of the more important colors would suffer." The firm tries first of all to make film that accurately records skin colors, then "common 'memory' colors such as those of the sky, grass, sand, etc." Chartreuse is less important. About manila the firm says nothing.

Everywhere along the creeks clumps of marshgrass fall from the steep banks into the channels, sinking to the grayish-black muddy bottom and waving their stems in the current. After several days the blades of grass, dying from too much immersion, turn chartreuse in color. At high tide, as *Essay* glides above them, they swirl like tropical anenomes. Uncharted and unphotographable, but chartreuse beyond doubt.

"The beauty that shimmers in the yellow afternoons of October—who ever could clutch it?" asks Emerson. What is shimmering? What is looming? What of the terms of light—not color—the visualist murmurs as *Essay* drifts? *Loom*, the 1914 Century explains, is "to appear indistinctly; come dimly into view, as from below the horizon or through a mist; rise up before the vision so as to give the impression of indistinct bulk or largeness; stand out prominently in the prospect." So loom is more than the inboard shaft of the oar, vastly more than the weaving machine. Melville begins *Moby-Dick* with a chapter called "Loomings," but who now knows the visual eeriness of the term, the mirage-like mix of shimmering and glimmering that casts color language into the old terminology of estuary English.

*

Essay visits the marshes, the vast realm so empty of other vessels. *Visit* comes from *visum* "to see," says the 1934 Webster. It and *vision* stand related. To visit means to see, not to talk, but to take notice, to take note, to actively engage the eye.

Gutters, guzzles, skiffs, and sea-marks, nothing particularly special, nothing particularly profound. Just the constituents of the vastness neither land nor sea, the place that changes with the tides, the place in which internal-combustion engine technology and the fresh-water canoe tradition fail alike, the place lacking sublimity, grandeur, and Coca-Cola machines. The place not attractive to anyone, not to the kids fascinated by the bikini-rigged beaches, not to the yachtsmen in debt for nylon sails and mini-radars, not to the Sierra Club types who prefer hiking or walking to slogging and splashing, not to anyone unequipped with the old language or unballasted with solid dictionaries.

Landscape — or seascape — that lacks vocabulary cannot be seen, cannot be accurately, usefully visited. It is not even theoretical, if *theory* means what the Greek root *theoria* means, a spectacle, a viewing. To be sure, in 1734 Jonathan Edwards, a Puritan theologian almost as intrigued by the natural world as by the hereafter, put his finger on the absolute necessity of first-hand experience. "There is a difference between having a rational judgment that honey is sweet, and having a sense of its sweetness," he explains in "A Divine and Supernatural Light." "A man may have the former, that knows not how honey tastes; but a man cannot have the latter unless he has an idea of the taste of honey in his mind." But how to provide the beginnings of a rational judgment about the marsh realm to those unable or unwilling to venture into it? The sea of green itself appears to entice very few casual visitors. After all, the hiker or bicyclist or automobile-imprisoned summer visitor finds no way in, no way permitting passage dry-shod, and most boaters shun the surf in the mouth and the creeks dried out at low tide. Without a local guide, someone to point out the different intensities of green marking creeks and guzzles perhaps, the foreigner may notice nothing. Contemporary society teaches a love of sandy beaches and ocean breakers, a dislike of brownish-grey (manila?) oozing mud and rank, un-lawn-like grass, and nothing much about grassland anyway, marsh or short-grass prairie or savannah. How many tourists visit national forests? How many notice a national grassland? Simply put, the marsh realm is neither landscape nor seascape.

How much ocean must intrude in the view or on the canvas before landscape becomes seascape? The 1935 *Thorndike-Century Junior Dictionary*, intended for children between the ages of ten and fifteen—the age when children start to relish exploring by themselves—defines *landscape* as "a view of the land" and "picture showing a land scene," but is silent about *seascape*, perhaps with good reason. The 1934 Webster explains *landscape* as "a portion of land or territory which the eye can comprehend in a single view, including all the objects so seen, esp. in its pictorial aspect," and "a picture representing inland natural scenery, as fields, hills, forests, water, etc." But the great dictionary directs its reader to definitions of two other words. *Seascape* it defines as "a view of or over the sea" and "a picture representing a scene at sea," and *marine*, among many definitions, as "a picture of some marine subject; a sea piece." What the 1934 Webster emphasizes is the complexity of *marine* versus *maritime*, noting that *marine* "designates esp. that which belongs to or is produced by the sea," while *maritime* "more often applies to that which borders on the sea." A fineness of discrimination about alongshore things suffuses all the definitions, making the reader aware that seascapes perhaps occupy a category between landscapes and marines.

Such discrimination is gone from the 1966 Random House, which defines *seascape* merely as "a view, sketch, painting, or photograph of the sea," the sort of truncated definition one expects to find in the *Thorndike-Century Junior Dictionary*. Something of alongshore discrimination is lacking.

More than the declining importance of coastwise commerce may explain the brevity of the Random House definition—and the ignorance of the editors of the Random House of any subtleties between *marine* and *maritime* too. Color definition offers an answer. The Random House defines *sea-green* only as "clear, light, bluish green." About *aqua*, which it defines as "a light, greenish-blue color," it says nothing in the definition of *sea-green*, suggesting something of the vagueness now informing dictionary definition of color—and of seacoast.

SEA-GREEN designates more than a color. A sea-green is a place, a place defined in the 1934 Webster as "land overflowed by the sea in spring tides." Years earlier, the *American Encyclopedic Dictionary* knew sea-greens as "ground overflowed by the sea in spring-tides," as did the Funk & Wagnalls *New Standard Dictionary*. What distinguishes land from ground, landings from fishing grounds, why the subtle difference between *land* in the Webster and *ground* in the other definitions? The marsh realm through which *Essay* wanders is not a sea-green. Tides regularly overflow the marshes, turning the whole place into a vast expanse of sea, interrupted only by sea-marks and occasional islands. Actual sea-greens lie just barely inland, just beyond the innermost wrack-lines, just beyond the last clumps of beach rose and the perfumeless sea lavender. There, in spring tides and hurricanes, the ocean intrudes, its seasonal swampings creating and maintaining odd wetland ecosystems. If the greenhouse effect proves true and the ocean level rises, then politicians, highway engineers, biologists, real-estate developers, and others will pay much attention to the sea-greens becoming saltmarsh. But who will speak of sea-greens by name?

*

Essay pokes and pulls and motors along, tracing the creeks and beaching near guzzles, probing the utterly common landscape of the salt marshes. And always, always sounding the depths of common speech, the vernacular alongshore speech fading from dictionary English.

About the Author

John R. Stilgoe is the Robert and Lois Orchard Professor in the History of Landscape at Harvard University, and the author of *Common Landscape of America, 1580–1845*; *Metropolitan Corridor: Railroads and the American Scene*; *Borderland: Origins of the American Suburb, 1820–1939*, and *Alongshore*.